T0296996

CAMBRIDGE LIBRARY COLLECTION
Books of enduring scholarly value

Mathematical Sciences

From its pre-historic roots in simple counting to the algorithms powering modern desktop computers, from the genius of Archimedes to the genius of Einstein, advances in mathematical understanding and numerical techniques have been directly responsible for creating the modern world as we know it. This series will provide a library of the most influential publications and writers on mathematics in its broadest sense. As such, it will show not only the deep roots from which modern science and technology have grown, but also the astonishing breadth of application of mathematical techniques in the humanities and social sciences, and in everyday life.

Atomicity and Quanta

Sir James Jeans (1877–1946) is regarded as one of the founders of British cosmology, and was the first to suggest (in 1928) the steady state theory, which assumes a continuous creation of matter in the universe. He made many major contributions over a wide area of mathematical physics, but was also well known as an accessible writer for the non-specialist. This is the full text of his Rouse Ball Lecture given in 1925 at Cambridge University, which surveyed the field of atomic and subatomic physics in the early days of quantum mechanics, with a brief historical perspective on measurement.

Cambridge University Press has long been a pioneer in the reissuing of out-of-print titles from its own backlist, producing digital reprints of books that are still sought after by scholars and students but could not be reprinted economically using traditional technology. The Cambridge Library Collection extends this activity to a wider range of books which are still of importance to researchers and professionals, either for the source material they contain, or as landmarks in the history of their academic discipline.

Drawing from the world-renowned collections in the Cambridge University Library, and guided by the advice of experts in each subject area, Cambridge University Press is using state-of-the-art scanning machines in its own Printing House to capture the content of each book selected for inclusion. The files are processed to give a consistently clear, crisp image, and the books finished to the high quality standard for which the Press is recognised around the world. The latest print-on-demand technology ensures that the books will remain available indefinitely, and that orders for single or multiple copies can quickly be supplied.

The Cambridge Library Collection will bring back to life books of enduring scholarly value (including out-of-copyright works originally issued by other publishers) across a wide range of disciplines in the humanities and social sciences and in science and technology.

Atomicity and Quanta

JAMES JEANS

CAMBRIDGE UNIVERSITY PRESS

Cambridge, New York, Melbourne, Madrid, Cape Town, Singapore,
São Paolo, Delhi, Dubai, Tokyo, Mexico City

Published in the United States of America by Cambridge University Press, New York

www.cambridge.org
Information on this title: www.cambridge.org/9781108005630

© in this compilation Cambridge University Press 2010

This edition first published 1926
This digitally printed version 2010

ISBN 978-1-108-00563-0 Paperback

ATOMICITY AND QUANTA

ATOMICITY AND QUANTA

BY

J. H. JEANS, D.Sc., LL.D, F.R.S.

Formerly Stokes Lecturer in Applied Mathematics
in the University of Cambridge; Sometime Professor
of Applied Mathematics in Princeton University

BEING THE ROUSE BALL LECTURE
DELIVERED ON MAY 11, 1925

CAMBRIDGE
AT THE UNIVERSITY PRESS
1926

CAMBRIDGE UNIVERSITY PRESS
Cambridge, New York, Melbourne, Madrid, Cape Town, Singapore, São Paulo, Delhi

Cambridge University Press
The Edinburgh Building, Cambridge CB2 8RU, UK

Published in the United States of America by Cambridge University Press, New York

www.cambridge.org
Information on this title: www.cambridge.org/9780521744812

First published 1926
This digitally printed version 2008

A catalogue record for this publication is available from the British Library

ISBN 978-0-521-74481-2 paperback

ATOMICITY AND QUANTA

TO our primitive ancestors measuring probably meant much the same thing as counting. The content of a flock of sheep was estimated merely by counting heads, and the length of a journey was recorded as being so many days' marches, the method being perhaps that of cutting a notch in a stick at the close of each day. It must soon have emerged that a measurement by integral numbers, while adequate for a flock of sheep, was quite unsuitable for the length of a journey, the reason, stated in modern scientific phraseology, being that sheep are "atomic" while a journey is not. At a quite early stage of civilisation, men must have appreciated the need for two distinct kinds of measurement —measurement by integers and measurement by continuously changing quantities. For a primitive people, all measurement of substance could be most readily expressed in terms of the space or area occupied, so that the two types of measurement reduce to counting and to the measuring of areas and volumes. Hence arise the two fundamental sciences of arithmetic and geometry; the shepherd was the primitive arithmetician, the land-surveyor and builder the primitive geometer.

Starting from such a basis it was natural that the earliest of thinkers should suppose that space was continuous, and that the objects that filled it were atomic. For space had always been the special province of the geometer, who had treated it as suited for continuous measurement, while objects in space had always provided occupation for the arithmetician, whose professional knowledge was concerned primarily with integers. The atomic theories of Leucippus, Democritus and Lucretius need not be thought of as brilliantly original conjectures; rather were they the natural development of thought along lines which were already traced out by practices dating from remote antiquity. Probably these thinkers deserve far more credit for ingenuity in the advocacy of their ideas than for any originality in the ideas themselves.

Granted, however, that it was natural, and perhaps inevitable, that space should be deemed continuous and matter atomic, what is to be said about time? Clerk Maxwell makes the statement, I know not on what authority, that up to the time of Zeno "time was still regarded as made up of a finite number of moments, while space was confessed to be divisible without limit." Jowett, it is true, attributes to Plato the statement that "time is to arithmetic what space is to geometry," but this appears to mean nothing more recondite than that time is one-dimensional, while space is three-dimensional. If the Greeks really regarded time as atomic—as a succession of moments—we have a simple explanation of the otherwise incomprehensible puerility of the famous paradoxes of Zeno. The best way of attributing any reasonable interpretation to these is perhaps to regard them as shewing that the hypotheses of non-atomic space and atomic time lead to absurdities contrary to experience.

What were matters of conjecture to the

Greeks are matters of reasonable certainty
to us to-day. The advent of the theory of
relativity has made it impossible to admit
that any such fundamental distinction as
that between continuity and atomicity can
exist between the properties of space and
time. If space is continuous, time must be
continuous also, for what is one man's
space is another man's time, and vice-
versa. The only type of atomicity which
relativity can admit as being theoretically
possible is one which finds its natural expression in terms of the four-dimensional
continuum, the continuum in which the
three dimensions of space and the one
dimension of time enter as four equal part-
ners. There can hardly be an atomicity of
the continuum itself, for, if there were, a
universal constant of the physical dimensions of space multiplied by time ought to
pervade the whole of physical science.
Nothing of the kind is even suspected,
nor, so far as I know, has ever been so
much as surmised. Thus science can to-day
proclaim with high confidence that both
space and time are continuous.

In the continuous framework provided by space and time the phenomena of Nature occur, or at least appear to us to occur, and our description of these phenomena centres round the two concepts of matter and of motion of matter, or of energy. We have a parallel arrangement of four terms:

Space —Time
Matter—Energy.

Relativity extends space into time, and the same extension extends matter into energy. The terms in the top line are, as we have seen, continuous. Of the terms in the lower line, one at least is known to be atomic. We believe matter to be constituted of indivisible atoms, no longer the atoms of the chemist which have long ago been found to be capable of division, but the fundamental electric charges of which all matter is supposed to be formed. If matter is atomic, can this atomicity be extended to energy in the same way in which space is extended into time, or what is the appropriate extension of the atomicity of matter?

The two terms in the top line of our scheme, although fundamentally similar in a way not yet fully understood, are measured in terms of different units. There is a "rate of exchange" between these two units just as there is between Italian lire and pounds sterling, and this rate of exchange, the ratio of a unit of length to a unit of time, is measured by the velocity of light, which we denote by C. If we wish to regard a second of time as a length we treat it as being equivalent to three hundred million metres, this being the distance travelled by a ray of light in one second. In the same way there is a rate of exchange between the two terms in the second line of our scheme, matter and energy, and this is known to be the square of the velocity of light, C^2. The addition of energy E to any material system increases its mass by a certain amount m such that $E = mC^2$; the addition can equally well be regarded either as an addition of mass or an addition of energy. As far back as 1881, Sir J. J. Thomson shewed this property to pertain to an electrically charged body, and it was, as we know, verified experimentally in the special case of a moving electron; Einstein's theory of relativity has now shewn it to be a property of all types of matter and of energy in general. Thus, if matter exhibits an atomicity of mass of unit m, the first thing we should naturally look for would be an atomicity of energy of unit mC^2.

Such an atomicity of energy almost certainly exists, and, although it does not shew itself in laboratory experiments, it is, I believe, of fundamental importance in the economics of the universe. As a matter of astronomical observation, young stars have far greater masses than old stars. By methods which need not be detailed here, it is possible to estimate the age-difference between any two stars or types of stars, and the star's total radiation of energy in this period of time just about represents the observed difference in mass provided the "rate of exchange" is taken to be C^2, the value demanded by the theory of relativity.

The long-standing puzzle of the source of
the sun's radiation appears at last to have
been solved, the answer to the puzzle being
the quite simple one that the source of the
sun's radiation is the sun's mass; the sun
keeps up its radiation by transforming its
mass into energy. It is a matter of indifference whether we say that the sun
radiates so many calories a minute or so
many tons a minute. If we select the latter
method of measurement the rate of radiation is about 250 million tons a minute,
and this represents a real decrease of the
sun's mass. Now if matter is essentially
atomic, the transformation of matter into
energy must also be atomic, and we are
brought face to face with the conception
of atoms of energy of amount mC^2, where
m is the atom of matter.

This atom of energy has been revealed
by astronomy alone; it has not so far come
within the purview of experimental physics.
This is hardly surprising; the interiors of
stars provide crucibles in which matter is
subjected to temperatures and pressures
thousands of times higher than the highest

that are attainable in our terrestrial laboratories. There is in any case no ground for surprise that the experiments which Nature conducts in her celestial laboratory reveal phenomena that cannot be repeated on earth, although in this actual case it seems likely that there is the still simpler explanation that the stars contain, and transform into energy, substances which are not available on earth at all—chemical elements of higher atomic weight than any known to us.

While astronomy has been unravelling the story of this simple atomicity of energy, experimental physics has discovered a humbler but far more complicated atomicity, of which it seems likely that the simple astronomical atomicity may be a limiting case.

About the end of the nineteenth century it became clear that the classical laws of mechanics predicted that all available energy should rapidly transform itself into radiation of infinitesimal wave-length. The prediction followed inevitably from the classical laws, so that if it was not confirmed by experiment, these laws had necessarily to be given up. A crucial test was provided by what is known as "cavity-radiation." A body with a cavity in its interior is heated up to incandescence; the radiation imprisoned in the cavity is let out through a small window and analysed in a spectroscope. If the classical laws were true the whole of this radiation ought to be found at the extreme ultra-violet end of the spectrum, independently of the temperature to which the body had been heated. In actual fact it is usually found to be mainly in the red parts of the spectrum, and under no circumstances does it ever conform to the predictions of the classical laws.

In 1900 Planck discovered the true law of spectral distribution of cavity-radiation by experimental methods, at the same time shewing how it could be deduced theoretically from a system of mechanics which differed widely from the classical system. He first assumed the radiation to be emitted by isochronous vibrators; the assumption that a vibrator of frequency ν could only possess energies $h\nu$, $2h\nu$, $3h\nu$, ..., and none other, used in conjunction with a certain amount of classical electrodynamics and thermodynamics, was then found to lead to the desired result. The assumption in question implied the existence of a unit of energy $h\nu$, but this was not a universal "atom" since its amount depended on the frequency ν of the vibrator with which it was associated. The multiplying factor h was, however, believed to be a universal constant of Nature, as has since proved to be the case. Planck's discovery of the existence of this constant in 1900 gave the first inkling of an atomicity which has since been found to pervade the whole of physical and chemical science. Some years later Poincaré (1912) and myself (1910) shewed that Planck's analysis could be reversed; from the circumstance that cavity-radiation was observed to conform to Planck's law, it could be deduced that the energy of the vibrators could only exist in integral multiples of $h\nu$, assuming that these vibrators were the true source of the emitted radiation.

Modern research, while confirming the accuracy of Planck's empirical law of cavity-radiation, has shewn the necessity for somewhat modifying his theoretical deduction of this law. Planck's original proof depended essentially on the supposed existence of isochronous vibrators, whose energy could only change by integral multiples of $h\nu$; the modern physicist discards the isochronous vibrators, which seem to have had no justification except as a scaffolding, and regards the radiant energy of frequency ν in space as changing its total amount only by integral multiples of a unit $h\nu$. The fundamental basis of the theory is now taken to be the law that emission and absorption of energy of frequency ν can only take place by whole units of amount $h\nu$. These energy-units $h\nu$ are the "quanta" from which the quantum-theory takes its name.

The law in the form just stated would suffice to lead to the observed law of cavity-radiation, but experimental physics adds further precision by shewing that the number of quanta involved in any single change is always unity. This is expressed by the law, commonly known as Einstein's law, that emission and absorption of radiation occur by single quanta $h\nu$, or, expressed as an equation,

$$E_1 - E_2 = \pm\, h\nu,$$

where ν is the frequency of the radiation concerned, and E_1, E_2 are the energies of the material system before and after the change.

The simplest illustration of this equation is provided by Einstein's photochemical law: "In a photochemical reaction the number of molecules which are affected is equal to the number of quanta of light which are absorbed." As a corollary to this law, the work done by one powerful quantum cannot be done by two weak quanta, or indeed by any number of weak quanta. If a quantum of a certain specified energy is needed to affect a molecule, quanta of lower energy, no matter how great their number, will produce no effect at all. The law not only prohibits the killing of two birds with one stone, but also the killing of one bird with two stones. Since the energy of a quantum $h\nu$ is proportional to its frequency ν we understand how a small amount of violet light can accomplish what no amount of red light would suffice to do, a fact familiar to every photographer—we can admit quite a lot of light if only it is red, but the smallest amount of violet light spoils our plates.

A similar illustration is provided by the photoelectric effect. Quanta more energetic than those requisite for photochemical action may suffice to break up the atoms on which they act, giving rise to the photoelectric effect. Again we find that there is a limiting frequency of radiation below which action does not take place no matter how intense the light. But as soon as the frequency passes this limit, even the feeblest intensity of light starts photoelectric action at once. Again it is found that the absorption of one quantum breaks up one atom, or rather ejects one electron from the atom; if the energy of the quantum is greater than that necessary to remove the electron from the atom, the excess is used in setting the electron into motion.

In this way we begin to see that what is atomic is not material energy at all, as Planck originally thought, but radiant energy, or to be more precise the exchanges of energy between radiation and matter. We begin to understand why the ultra-violet end of the spectrum of a hot body shews so little energy; the simple reason is that very, very few of the molecules or atoms in a hot body possess enough energy to emit a complete quantum of violet radiation. In a hot body at temperature T, the vast majority of atoms possess energy of amount comparable with RT, where R is the absolute gas-constant. The proportion of atoms whose energy is a large multiple of this, say nRT, where n is a large number, is of the order of e^{-n} (or, strictly, its logarithm to base e is of the order of $-n$). Thus, when n is large, this proportion is very small indeed. If the quantum of radiation of any frequency ν is equal to nRT, where n is large, the proportion of atoms which are in a position to shed a whole quantum $h\nu$ of radiation is a small quantity of the order of e^{-n} or $e^{-h\nu/RT}$, whence it follows that the amount of radiation of high frequency ν is also a small quantity of the order of $e^{-h\nu/RT}$.

According to the classical mechanics, the exchange of energy between matter and radiation was a continuous process, there being no limit to the smallness of the amount of energy that could be exchanged. To represent exchanges of energy as envisaged by the classical mechanics we must put $h = 0$ in the quantum law, and in actual fact if we put $h = 0$ in Planck's law of cavity-radiation

$$\frac{8\pi}{c^3} \frac{h\nu}{e^{h\nu/RT} - 1} \nu^2 d\nu,$$

it reduces to the law of radiation

$$\frac{8\pi RT}{c^3} \nu^2 d\nu,$$

which was predicted by the classical mechanics.

Planck's fundamental discovery can be stated in the form that h is different from zero, and it is this concept that forms the corner-stone of the quantum-theory. The value of h is so small that it had been over-looked by the classical mechanics, and indeed the effect it produces is only appreciable in phenomena which are of the atomic scale of smallness and in phenomena, such as cavity-radiation, which arise directly out of phenomena on the atomic scale.

Small though the value of h is, we must recognise that it is responsible for keeping the universe alive. If h were strictly zero the whole material energy of the universe would disappear into radiation in a time which would be measured in thousand-millionth parts of a second. For instance, the normal hydrogen atom, owing to continuous emission of radiation, would begin to shrink at the rate of over a metre a second, and after about 10^{-10} seconds the nucleus and electron would fall together and would probably disappear in a flash of radiation. The quantum-theory, by prohibiting any emission of radiation less than $h\nu$, prohibits in actual fact any emission at all except from those few atoms which have a quite exceptionally large amount of energy to emit, whence it results that the duration of the universe, instead of being measured in units of 10^{-10} seconds, is measured in units of 10^{20} seconds. A parallel case would occur if the legislature of a country, deploring the amount of money its inhabitants spent on beer, decreed that no beer should be bought or sold except by units of a thousand gallons at a time, and further enacted that one purchaser alone should be concerned in one purchase. The method would obviously be effective, although perhaps not popular; Nature, however, is not dependent on a democratic vote.

Following a line of thought brilliantly opened up by Bohr, the quantum-theory has shewn that the internal arrangements in matter are, in a sense, atomic. The hydrogen atom consists of an electron describing a circular or elliptic orbit about a positive nucleus, but only certain orbits are possible and these form only an infinitesimal fraction of the orbits which would have been possible under the classical electrodynamics. For instance no orbits are possible unless their angular momentum is an integral multiple of a certain "atom" of angular momentum, the amount of the "atom" being $h/2\pi$, where h is Planck's universal constant. And no elliptic orbits are possible unless the ratio of the axes of the ellipse (b/a) is a commensurable number. The radii of the possible circular orbits are proportional to the squares of the integral numbers 1^2, 2^2, 3^2, 4^2, ..., while the energies are inversely proportional to these squares. The semi-major axes of the various elliptic orbits which are possible are equal to the radii of the possible circular orbits, so that the energies also are equal to the energies of the corresponding circular orbits. A similar type of atomicity governs the internal dispositions of more complex atoms.

In place of a gradual shrinkage of the energy of an atom, such as is predicted by the classical mechanics, the quantum-theory introduces the conception of definite "energy-levels," and changes of energy result from the passage from one energy-level to another. An atom can stay for a very long time with its energy constant, and equal to the value corresponding to one of these energy-levels, and during this period it emits no radiation. At intervals it may jump from one energy-level to another, and it is at these jumps only that energy is emitted or absorbed. A hydrogen atom losing energy by radiation must be compared, not to a ball rolling downhill, but rather to a ball bouncing down a flight of steps.

This conception brings a new meaning into the fundamental equation

$$E_1 - E_2 = \pm\, h\nu.$$

The student whose approach to this equation is through the classical electrodynamics is apt to think of ν as the fundamental datum, the frequency of an oscillation or of the free vibrations of an oscillator, and of the left-hand side as expressing the corresponding change of energy. But it now appears that the student of classical electrodynamics would have got hold of the wrong end of the stick; it is the terms on the left which constitute the fundamental data of the problem; both E_1 and E_2 being immutably fixed by the structure of the atom itself, while ν is the variable which adjusts itself to conform to the given values of E_1 and E_2. We used to interpret the frequency of light as meaning that the ether was set into rhythmic vibrations by the occurrence of isochronous oscillations of the same frequency in matter. We know now that such oscillations do not exist. An electron suddenly stumbles in its orbit and falls to a lower energy-level. Radiant energy is set free as a consequence and it is merely the amount of this energy which determines its frequency: the radiation chooses a wave-length such that its amount shall be equal precisely to one quantum.

This result is so entirely unexpected, and effects such a complete reversal of what the classical electrodynamics had led us to anticipate, that it is natural to pause to inquire whether ν, in our fundamental equation, is really a frequency at all in the old-fashioned sense. The answer, and a very complete answer, is provided by letting some of the light whose frequency is in question fall on a diffraction grating. It is at once found that the frequency predicted by our fundamental equation agrees exactly with the wave-length indicated by the grating: indeed, it is by innumerable agreements of precisely this type that the quantum-theory has won general acceptance.

We have now seen how radiant energy is emitted in complete quanta, the wave-length of the radiation adjusting itself to such a value that the quanta shall be complete. But this does not require that the whole radiant energy in space should exist only in complete quanta, for we have so far found no reason for thinking that the quanta retain their individual entities after the process of emission has taken place. Obviously there are two possibilities open to the quanta when once they are free in space:

(A) The energy, once freed from matter, may spread as directed by the classical laws of electrodynamics (Maxwell's equations), so that the quanta lose their individuality immediately after their birth.

(B) The energy may travel in complete quanta like bullets fired from a rifle, each quantum retaining its individual identity from the moment of its emission until it is again absorbed by matter.

It might seem to be an easy matter to decide between two such entirely conflicting possibilities, but this has not proved to be the case, and many physicists still regard the question as an open one. At first sight most phenomena arising from the emission of radiation seem to point to possibility (A), while those arising from its absorption seem to point with equal cogency to possibility (B).

An instance of the latter type of phenomenon is provided by the X-ray photoelectric effect. Radiation which breaks up atoms is absorbed in complete quanta, one for each atom, any superfluous energy being used in endowing the liberated electron with kinetic energy. The quantum of X-radiation possesses an enormous amount of energy, as a consequence of the high frequency of the radiation, so that when a quantum breaks up an atom, only a tiny

fraction of its total energy is spent in the
actual breaking-up process, and practically the whole reappears as the kinetic
energy of the electron. A stream of electrons may fall on to an anti-cathode, each
electron, as its motion is checked, emitting
a quantum of X-radiation. After travelling
through space, this X-radiation may break
up any atom it encounters and each electron ejected in this way will travel with
very approximately the velocity of the
original stream of electrons. A being who
had no knowledge either of radiation or
of quanta, might conjecture that each electron in the original stream had in some
way passed on its velocity across the inter-
vening space to some other electron, just
as in croquet the momentum of the mallet
is transmitted through the held ball to the
ball with which it is in contact. Knowing
what we know of the atomicity of quanta,
it is almost inevitable that our first conjecture should be that the energy of the
original electron had been transmitted in
the form of a quantum of radiation which,
travelling like a rifle-bullet through space,

had disrupted the first atom it struck and yielded up its whole energy to this atom.

If, however, radiation is to be compared to rifle-bullets, we know both the number and size of these bullets. We know, for instance, how much energy there is in a cubic centimetre of bright sunlight, and if this energy is the aggregate of energies of individual quanta, we know the energy of each quantum (since we know the frequency of the light) and so can calculate the number of quanta in the cubic centimetre. The number is found to be about ten millions. By a similar calculation it is found that the light from a sixth magnitude star comprises only about one quantum per cubic metre, and the light from a sixteenth magnitude star, only about one quantum per ten thousand cubic metres.

Thus, if light travels in indivisible quanta like bullets, the quanta from a sixteenth magnitude star can only enter a terrestrial telescope at comparatively rare intervals, and it will be exceedingly rare for two or more quanta to be inside the telescope at the same time. A telescope of double the aperture ought to trap the quanta four times as frequently, but there should be no other difference.

This, as Lorentz pointed out in 1906, is quite at variance with our everyday experience. When the light of a star passes through a telescope and impresses an image on a photographic plate, this image is not confined to a single molecule or to a close cluster of molecules as it would be if individual quanta left their marks like bullets on a target. An elaborate and extensive diffraction pattern is formed; the intensity of the pattern depends on the number of quanta, but its design depends on the diameter and also on the shape of the object-glass. Moreover the design does not bear any resemblance whatever to the "trial and error" design which is observed on a target battered by bullets. It seems impossible to reconcile this with the hypothesis that quanta travel like bullets directly from one atom of the star to one molecule of the photographic plate. Professor G. I. Taylor tested the hypothesis by photographing a diffraction pattern by very feeble light. Even when two thousand

hours of exposure were required the diffraction pattern remained.

Wien has recently studied the rate of
decay of radiation from matter. As an
instance, he finds that the characteristic
red radiation from hydrogen (H_α) is re-
duced to $1/e$ times its intensity in $\dfrac{1}{5\cdot 4 \times 10^7}$
seconds. If the radiation consisted of dis-
crete quanta, the most obvious interpretation of this result would be that the intensity of radiation in each quantum fell off
according to the same law, so that a quantum would necessarily be several metres in
length. It is true that an alternative explanation could be obtained by supposing
that the chance of a hydrogen atom emitting a bullet-like quantum falls off at the
rate in question. But with these experiments we may combine others which shew
that it is possible to obtain interference
over a path-difference of about a million
wave-lengths. This can hardly be attributed to bullet-like quanta, for the same
atom cannot emit two quanta of the same
radiation in succession, and it is difficult to

imagine that definite phase relations can exist between quanta emitted by independent atoms. Thus, the observed limits of interference suggest that quanta must be at least a million wave-lengths (say two feet) in length, while Wien's experiments make it highly probable that the length would be several metres. In any case the lengths which have to be attributed to quanta cannot be reconciled with common experience. If quanta were either feet or metres in length, they would not be indivisible. Every click of a camera shutter would cut several in halves, while Fizeau's toothed-wheel experiment for determining the velocity of light would represent a continuous massacre of quanta.

These instances, as well as others that could be quoted, shew the impossibility of reconciling the hypothesis of indivisible quanta with the phenomena of everyday life and in particular with interference phenomena. Interference might possibly be interpreted as a statistical effect if the statistical units could be assumed to be sufficiently small, but as things stand, a

statistical explanation is ruled out by the
known magnitude of the quantum. Whatever happens, it appears to be necessary
to retain the main features of the undulatory theory of light.

As it appears to be equally necessary to
retain the fundamental quantum equation

$$E_2 - E_1 = h\nu,$$

it becomes of primary importance to in-
vestigate how far this equation is com-
patible with the undulatory theory of light.
In fields, if any, in which the two become
incompatible, we must limit the application of one or the other. Before discussing
this question at length, it may be well to
remark that it only forms part of a wider
problem. Any solution of the problem of
the nature of radiation and of its interaction with matter must, I think, conform
to the four following general principles:

I. The quantum-principle, as expressed
by the equation $E_2 - E_1 = h\nu$.

II. The principle of the undulatory
theory of light.

III. The principle of relativity.

IV. The principle of reversibility.

The first two of these principles have already been explained, the third probably needs neither explanation nor justification at the present day, but the fourth, the principle of reversibility, may call for a few words of explanation.

The classical dynamics predicted that there could be no final steady state of equilibrium between matter and radiation until the whole energy of the matter had been transformed into ultra-violet radiation of infinitesimal wave-length. Thus, a hot-body cavity and the radiation imprisoned in it could never be in thermodynamical equilibrium. Planck's derivation of his law of cavity-radiation ran contrary to the classical dynamics in that it was based on the supposition that thermodynamical equilibrium existed, and whatever parts of Planck's original theory may have been found to need modification, this part still stands. Thus, all the processes which are taking place in a hot-body cavity in equilibrium with its own radiation must be thermodynamically reversible.

Let us turn now to a consideration of

how far, if at all, these four principles
appear to be in any way inconsistent. We
may suitably start with the first two, the
quantum-principle and the undulatory
theory. A great deal has been written
about the apparent inconsistency of these
two principles, but it may be remarked
that consistency can be secured from the
outset by limiting them to non-overlapping
fields. The quantum-principle makes no
claim to apply except to the interchange
of energy between matter and radiation—
indeed, it is meaningless except in this connection—whereas the undulatory theory,
in the form in which we have found it to
be indispensable, has no application except to radiation travelling freely in space
without interchanging energy with matter.
There need, then, be no inconsistency
between these two principles unless we
ourselves create it by permitting one principle to invade the domain of the other.

An instance of the way in which the
two principles can be used in conjunction,
each in its own domain, is provided by
the X-ray photoelectric effect already

described. In the first stage of this phenomenon a rapidly-moving electron hits a
target and is brought to rest. If we knew
the precise specification of the forces at
work on the electron, we could calculate
its deceleration at every instant and hence
the amount and spectral distribution of the
emitted radiation. But the undulatory
theory has no capacity for specifying
either these forces or the resulting radiation; we are here in the province of the
quantum-theory. This tells us that the
emitted radiation will be one quantum of
monochromatic radiation of a frequency
ν which can be immediately calculated
from the equation

$$\tfrac{1}{2}mv^2 = h\nu.$$

The quantum-theory can tell us nothing
as to what happens to this radiation after
it is born; here the undulatory theory
takes up the story and explains that the
radiation spreads through space with an
ever-expanding wave-front; it travels with
the velocity of light and with unchanged
frequency ν. Thus the stoppage of a

stream of electrons all of which had initially the same velocity v fills space with radiation of frequency ν. The quantum-theory resumes its sway as soon as any of this radiation again encounters matter and shews that each transfer of energy between the radiation and matter must be of amount $h\nu$ or $\frac{1}{2}mv^2$, so that any electron which is ejected must necessarily travel with a speed equal to that of the electrons in the original stream. The weak link in the chain is that neither theory has so far explained the mechanism which results in a complete quantum $h\nu$ of monochromatic radiation being concentrated in space in such a way that it can be swallowed at one gulp by an absorbing atom. We shall return to this problem later on.

Assuming that the quantum-principle and the undulatory theory can be brought into harmony by the simple device of restricting each to its own domain, let us examine whether either principle is incompatible with our third principle, the principle of relativity. No physicist will doubt that this principle must prevail in both domains, so that it must agree with each of the two principles already discussed, each in its own domain. It is already in agreement with the undulatory theory, since the general equation of wave-propagation conforms to the relativity condition, so that it remains only to examine how agreement can be secured between the relativity-principle and the fundamental quantum-equation.

Consider first the simple problem of the interaction of a field of radiation with a single free electron. The quantum-principle requires that any interchange of energy shall be of amount $h\nu$, where ν is the frequency of the radiation emitted or absorbed. Thus if v is the velocity of the electron before the absorption of a quantum and v' its velocity afterwards, the quantum-equation assumes the form

$$\tfrac{1}{2}mv^2 + h\nu = \tfrac{1}{2}mv'^2;$$

the sign of $h\nu$ must of course be changed for emission.

The principle of relativity requires that this relation shall remain true when the velocities and frequency are measured by an observer who is moving with any uniform velocity we please relative to the original observer. Let our second observer move with such a velocity that the electron appears to him to be at rest after the absorption of the quantum of energy. Then, $v' = 0$ and the equation assumes the form

$$\tfrac{1}{2}mv^2 + h\nu = 0.$$

Since neither $\tfrac{1}{2}mv^2$ nor $h\nu$, both of which represent positive energies, can be negative, the equation can only be satisfied if $v = 0$ and $\nu = 0$. We see that a free electron can absorb no radiant energy except that of zero frequency, this of course representing the energy of an electrostatic field.

To examine the possibility of emission of energy we change the sign of $h\nu$, and select axes such that $v = 0$. The equation now requires that $v' = 0$ and $\nu = 0$, so that there can be no emission of radiation except of frequency $\nu = 0$.

Such are the results obtained by harnessing the quantum-theory to the theory of relativity; let us examine what they imply. According to the classical theory, the passage of a wave of radiation of frequency $p/2\pi$ produces an electric force of amount

$$X_p \cos pt,$$

and this in turn produces an acceleration in a free electron equal to e/m times the force. According to the quantum-theory, as we have just seen, there can be no transfer of energy between such a wave of radiation and a free electron, so that the velocity of the free electron remains unaffected. In other words, an electric force X, if it results from a field of radiation, produces no acceleration in an electron, although we know that if it is part of an electrostatic field, as for instance in the case of cathode rays, it produces an acceleration Xe/m.

We begin here to get some conception of the enormous gulf which separates the quantum-dynamics from the classical theory. On the classical theory, electric force is usually defined by the general statement that an electric force X produces a mechanical force Xe, whereas according to the quantum-theory this property of electric force appears only to be true under very special conditions. Even on the quantum-theory, the field of force X, Y, Z must, as we have seen, be supposed to exist, but its properties are almost unknown, except that when the field is permanent its properties reduce to those assigned to it by the classical theory.

The mathematician may claim to have detected an inconsistency in the statement that an electrostatic field produces an acceleration in a free electron, while a field of radiation does not. He will say that a field of force is electrostatic only if it exists unaltered throughout all time, from $t = -\infty$ to $t = +\infty$. If this is not the true definition of an electrostatic field he will protest that he is unable to assign any precise meaning to the term. If an electric field varies ever so slightly, the value of the electric force at any point can be expressed, by Fourier's theorem, in the form

$$X = \int X_p \cos\left(pt - \epsilon_p\right) dp,$$

and the only part of the field which can
be called electrostatic is that which arises
from an infinitesimal range of values close
to $p = 0$. In other words, if a field vary
ever so slightly, the truly electrostatic part
is of only infinitesimal intensity.

The answer to this objection is, I think,
as follows. If we are going to take the
whole of time within our purview, we must
also take the whole of space, and when
this is done, no electron can be regarded
as strictly free. Either we may limit ourselves to a small region in space and time,
in which we can have both free electrons
and an electrostatic field of finite intensity,
or we can survey the whole of space and
time, in which case no electron is strictly
free and our theorem becomes valueless.

A further complication, however, remains. Consider the case of a hypotheti-
cal universe containing only two material
constituents, a negative electron and a
positive proton of equal charge situated at,
let us say, a distance of a million light
years apart. Does such a universe consist
of two free electric charges, or of a hydro-

gen atom, or is it a matter of indifference in which way we regard it? The last alternative can be dismissed at once, for two free charges cannot absorb or emit radiant energy, whereas a hydrogen atom can. We see at once that a hydrogen atom must be something more than the sum of its material constituent parts. In a sense we knew this already, for two electric charges, so long as they are regarded as independent entities, can be at any distance apart and can have any velocity relative to one another, but the combination only forms a hydrogen atom when these distances and velocities satisfy the very definite conditions discovered by Bohr and Sommerfeld. We now see that the "something more" which must be added to an electron and a proton to form a hydrogen atom is the mechanism which enables the structure to absorb and emit radiation.

When the electron and proton in a hydrogen atom are so far apart that both the potential and kinetic energy of the electron may be neglected, the absorption of a quantum of radiation of frequency ν endows the electron with a velocity v such that

$$\tfrac{1}{2}mv^2 = h\nu.$$

This is a direct consequence of the quantum condition, our equation being the limit of the ordinary photoelectric equation when the energy of ionisation is left out of account. The electron acquires momentum in the direction of its newly-acquired velocity equal to mv or

$$\frac{h\nu}{\tfrac{1}{2}v}.$$

Now even if the whole of the absorbed radiation $h\nu$ had been moving originally in precisely the direction of the velocity v, the momentum of this radiation would only have been

$$\frac{h\nu}{C},$$

a quantity which is exceedingly small in comparison with the momentum acquired by the electron. The proton must acquire a backward momentum equal to or greater than

$$\frac{h\nu}{\frac{1}{2}v} - \frac{h\nu}{C},$$

and this is practically equal to the forward momentum of the electron. We may picture the process of photoelectric ejection by thinking of the nucleus and electron as forming a rifle and a bullet respectively. The quantum of energy which is to be absorbed acts as a charge of gunpowder, which, while contributing very little momentum itself, endows the rifle and bullet with equal and opposite momenta, while practically the whole of the energy goes into the bullet. This shews very clearly that absorption is not solely by the electron, to which the visible effect is confined, but by the atom as a corporate whole. Some part of this corporate whole, the rifle-barrel, must perform the double function of trapping the radiant energy and distributing its energy and momenta between the two constituents of the atom. What is this mechanism?

The conjecture which first presents itself is that it may be an actual "tube of

force" connecting the two material constituents of the atom. A free electron is
surrounded by a spherical field of force
which may be represented by lines of force
radiating in all directions equally from its
surface. We are tempted to imagine that
when the electron is bound in an atom,
all these lines of force may be gathered up
into a sort of rope binding the electron to
the nucleus of the atom. The picture is an
attractive one, but we must be careful not
to regard it as anything more than a picture.
In actual fact there is abundant evidence
that the field of force from a nucleus is
not bunched up into a number of tubes
which pass to the surrounding electrons.
Experiments on the scattering of β-par-
ticles shew that when a free electron enters
an atom, the whole force Ne/r^2 from the
nucleus is available to act, and does act,
upon the wandering electron. Whatever
connection we imagine between the nucleus and the electrons which are "bound"
to it, there must be no tampering with the
full field of force of the nucleus. Moreover, the picture of discrete tubes of force

leaves no room for the structure of molecules, and so would crush the science of chemistry out of existence. For when each atom of hydrogen becomes a self-contained system, there is no reason why these atoms should associate themselves into molecules.

Nevertheless, the picture of concrete tubes of force can render one service to the quantum-theory. If tubes of force had a real existence, the strength of the tube connecting a nucleus to an electron would have the universal value $4\pi e$; these tubes would be atomic, their atomicity being of course derived from the fundamental atomicity of electric charges. Now the value of h is found to be given by an equation of the form

$$\frac{hC}{2\pi} = K\,(4\pi e)^2,$$

where K is a constant which is not greatly different from unity. This suggests that the atomicity of h may arise from the atomicity of $4\pi e$, the strengths of the tubes of force. No doubt we must take a wider view, but it seems possible that the quantum-atomicity may in the long run prove to be nothing but another aspect of the atomicity in Nature which limits electric charges to being multiples of the fundamental charge e.

Let us now turn to the fourth principle, that of reversibility, and examine how this can be brought into accord with the first three principles. There is no conflict between this principle and the principle of relativity; neither is there any conflict between it and the quantum-equation

$$E_1 - E_2 = h\nu,$$

for E_1 and E_2, the energies before and after a change, enter symmetrically except for the matter of algebraic sign which represents the difference between emission and absorption. Further, there is no fundamental conflict between the principle of reversibility and the undulatory theory of light, since the wave-equation

$$\left(\frac{\partial^2}{\partial x^2} + \frac{\partial^2}{\partial y^2} + \frac{\partial^2}{\partial z^2}\right)\phi = \frac{1}{C^2}\frac{d^2\phi}{dt^2}$$

conforms to the principle of reversibility, retaining its form when the sign of dt is

changed. There is however a conflict between the principle of reversibility and those
solutions of the wave-equation which are
usually adopted in the classical electrodynamics, and this conflict the quantum-
dynamics must contrive to avoid. A simple
instance may bring the problem into focus.

In 1901 Professor E. P. Adams attached
a series of charged brass spheres to the
circumference of a rotating wheel and
found that these produced a magnetic field
which alternated periodically as the spheres
passed by a suspended magnetic needle.
The electric field of course also varied
periodically, and on calculating the Poynting Flux we can estimate the rate at
which energy was radiated away from the
rotating system. No physicist is likely to
doubt that this radiation actually took
place; in all essentials the method of its
production was similar to that by which a
field of radiation is generated in radio-
telegraphy, and there can be no doubt as to
the reality of the radiation here. Further,
no physicist will doubt that if the number
of spheres had been reduced to one of

negative charge, and if a positive charge of equal amount had been placed at the centre of the circle, an emission of radiation would still have occurred. Yet such a pair of charges attached to a revolving wheel forms a realistic large-scale model of the hydrogen atom. How then comes it that the spheres radiate energy while the hydrogen atom does not?

There is of course an enormous difference of scale, but this is easily allowed for and we find that if the hydrogen atom radiated according to the same laws as the revolving spheres, its life would be limited to a small fraction of a millionth of a second. Clearly we must look for something more fundamental than differences of scale. The essential difference, as we shall see, is that the process studied by Professor Adams was irreversible, while the processes in the normal hydrogen atom are strictly reversible.

We know that heat applied to one point of a body tends to spread through the body until a uniform temperature is attained. The phenomenon can be readily explained

in terms of the second law of thermodynamics. A configuration in which an
undue amount of heat is concentrated
around one spot is a configuration of low
entropy. There can be no equilibrium
until the entropy has attained a maximum
value, and this only occurs when the heat
is evenly distributed throughout the mass.
Similarly, on the undulatory theory of
radiation, a configuration in which radiation is concentrated unduly in one spot is
one of low entropy and in striving to
reach maximum entropy the radiation
tends to spread throughout the whole of
space. Adams' spheres radiated away their
energy because by so doing they increased
the entropy of the whole system, but this
in itself shews that their radiation was,
thermodynamically, an irreversible process.

On the other hand, when a system of
hydrogen atoms is in thermodynamical
equilibrium with radiation, there can be
no further increase of entropy and so
no spreading out of spherical waves, unless indeed there is a corresponding

convergence of spherical waves onto a
small region.

The general solution of the wave-equation, which is symmetrical about the origin
and so represents spherical waves, is

$$\phi = \frac{1}{r}\left\{F\left(t - \frac{r}{a}\right) + \Phi\left(t + \frac{r}{a}\right)\right\},$$

where F and Φ are any functions whatever. In the development of classical electrodynamics, it has been usual to reject the
second term in this solution altogether on
the ground that it represents convergent
spherical waves such as do not occur in
Nature. This limitation to divergent waves,
however, necessarily excludes any application to systems in thermodynamical equilibrium with radiation, such as normal
hydrogen atoms. Such systems must be
capable of running backwards as well as
forwards, whereas the divergent waves represented by the first term in ϕ can only
run forwards.

To obtain a solution which is capable of
representing reversible processes, we must
give equal prominence to convergent and

divergent waves, and so must take Φ to
denote the same function as F. The value
of ϕ near the origin $r = 0$ now approxi-
mates to $2F(t)/r$, and the function F is
completely determined. In the usual nota-
tion for "retarded" potentials, we write
$[f]$ for $f(t - r/a)$; if we similarly denote
$f(t + r/a)$ by $\{f\}$, then the electromagnetic
potentials from a system of moving electric
charges which are in thermodynamical
equilibrium with radiation are found to be

$$\Psi = \iiint \frac{[\rho] + \{\rho\}}{2r} \, dx\, dy\, dz,$$

$$F = \iiint \frac{[\rho U] + \{\rho U\}}{2r} \, dx\, dy\, dz, \text{ etc.}$$

From these formulae we can calculate in
succession the electric and magnetic forces,
the Poynting Flux and the rate of radiation of energy. The latter, averaged over
a sufficient time, is found to be zero; there
is an ebb and flow of energy but no steady
streaming away from the system of moving
electric charges.

Thus even under the laws of classical
electromagnetism, it is possible for an atom

to be a permanent non-radiating electrical system. It is true that such an atom does not consist merely of electrical charges, but of charges plus a surrounding electro-magnetic field. This latter provides just the "something extra" which, as we have already seen, must be added to a system of electric charges before they can function as an atom.

It hardly seems possible that the model we have just imagined can present us with the true picture of an atom. There is the obvious objection that the electromagnetic field in which we have supposed the electric charges to be immersed not only extends to infinity but is of infinite energy. Not only so, but the model provides no explanation of the atomicity which spectroscopy shews to exist in the angular momenta, radii and eccentricities of atomic orbits. Before we can adopt this electromagnetic field as the missing part of our atom, we must give it some sort of a structure which will not only limit its total energy but will impress some sort of atomicity on the orbits which are possible for

an electron. As has already been seen,
there is no reason why this new atomicity
should not be precisely that which imposes
the atomicity of charge e on electrons and
protons, for the new atomicity involves no
physical constants beyond e and C.

So far as we can at present conjecture,
the investigation of the structure which
produces this atomicity appears to be the
big problem in the path of the quantum-
theory. To conform to the principle of
relativity, the new atomicity must admit
of expression in terms of the space-time
continuum, although we have seen that it
cannot be an atomicity of the continuum
itself. It may conceivably be an atomicity
of its metric properties, such as determine
its curvatures. We may perhaps form a
very rude picture of it by imagining the
curvature of the continuum in the neighbourhood of an atom not to be of the continuous nature imagined by Weyl, but to
occur in finite chunks—a straight piece,
than a sudden bend, then another straight
bit, and so on. A small bit of the continuum viewed through a five-dimensional

microscope might look rather like a cubist picture; and, conversely, perhaps a cubist picture looks rather more like a little fragment of the continuum than like anything else.

Since free electrons and protons, as we have seen, can neither emit nor absorb radiation, the mechanism of both absorption and emission must reside in the electromagnetic field which binds the electrons and protons together into an atom. In the Compton effect, parallel beams of X-radiation fall on matter; the emergent radiation is found to be no longer parallel and to be somewhat softer than the incident radiation. The phenomenon is generally regarded as one of scattering, and Compton has shewn that it admits of a simple explanation in terms of the bullet-theory of radiation. Bullets of radiation are supposed to hit electrons and to glance off at varying angles; part of their energy is imparted to electrons while the remainder passes on as a quantum of diminished energy, and therefore as a bullet of softer radiation. The supposition that the impact between the bullets of radiation and the electrons conforms to the laws of conservation of energy and momentum leads to the formula

$$\delta\lambda = \frac{h}{mC}(1 - \cos\theta),$$

for the increase in the wave-length λ of radiation which has been deflected through an angle θ.

If however we discard the bullet-theory of radiation, the phenomenon admits of almost as simple an explanation. The atom of which the electron forms part must first be supposed to absorb a quantum $h\nu$ of radiation in the ordinary way. This would normally set the electron free, endow it with a certain amount of momentum and produce a corresponding back-momentum in the remainder of the atom. We have next to suppose that in some way the process is checked and that before it is complete the atom emits a new quantum of energy $h\nu'$. If it can be supposed that the nucleus and remainder of the atom remain unmoved by the double process, then we obtain equations which would be identical with those leading to Compton's formula except in one respect. Compton's formula is derived on the supposition that each quantum of scattered radiation moves, bullet-like, along a definite line in space, whereas if the bullet-hypothesis is rejected, the emitted quantum of radiation must be free to spread in space. On account of this spreading, λ will not be the wave-length of the radiation which travels in the direction θ, but the mean wave-length, the radiation no longer being strictly mono-chromatic. A small change must also be made in the numerical coefficient h/mC. Except for these changes, the proposed explanation leads to the same formula as the bullet-theory, and it can hardly be claimed that experiments are as yet pre-cise enough to decide between the two formulae. The explanation just suggested may seem somewhat artificial, but it can claim to be far less at variance with the other phenomena of radiation than the usual explanation which postulates a bullet-like structure of radiation.

Another phenomenon which certainly appears at first to favour the bullet-theory is the photoelectric effect; we have already noticed the difficulty of supposing radiation to be absorbed by whole quanta unless there is some mechanism by which the radiation is bunched into quanta ready for absorption. Still, the photoelectric effect does not seem to be in any way incompatible with the undulatory theory, provided we are free to depart from the ordinary electromagnetic equations in the immediate neighbourhood of matter. In a field in which only two atoms exist, there is a solution of the classical electromagnetic equations in the form

$$\Psi = \iiint \frac{[\rho]}{r}\,dx\,dy\,dz + \iiint \frac{\{\rho'\}}{r'}\,dx'\,dy'\,dz',$$

$$F = \iiint \frac{[\rho U]}{r}\,dx\,dy\,dz$$
$$+ \iiint \frac{\{\rho' U'\}}{r'}\,dx'\,dy'\,dz',\ \text{etc.,}$$

in which the first integral in each equation, containing the "retarded" terms $[\rho]$,

$[\rho U]$, ..., refers solely to the first atom, while the second integral, containing the "advanced" terms $\{\rho'\}$, $\{\rho' U'\}$, ..., refers solely to the second atom. If the two atoms are of precisely similar structure, similarly oriented, and describing similar motions, the solution written down gives an average flux of energy equal to zero over a sphere surrounding both systems, so that radiation merely flows from one system to the other. At first the radiation spreads out from the emitting system as though the second system were non-existent, but is gradually drawn in towards this second system and finally concentrates wholly upon it in the form of a convergent wave. The flow of energy is analogous to hydrodynamical flow in a field in which there is a source and a sink of equal strength.

This illustration brings us to the very heart of the supposed inconsistency between the undulatory theory and the quantum-theory. On most of the occasions on which a really wrong turning has been taken by physical science, the mistake has

been made by allowing a faulty analogy
to take the place of exact mathematical
analysis. Newton's corpuscular theory of
light provides a conspicuous instance. By
analogy with what was observed in the
flight of projectiles it was supposed that a
crowd of corpuscles could be projected in
a parallel beam, and by analogy with what
was observed on the rippled surface of
water it was supposed that an undulatory
disturbance could not be so projected. A
full mathematical analysis would have
shewn that the last analogy failed when
the smallness of the wave-length of the
light-undulations was taken into account,
and, in actual fact, as soon as this analysis
was forthcoming, the corpuscular theory
fell. And again to-day the analogy of
flying bullets, an analogy within the experience of every one, is ready to hand to
suggest explanations for the photoelectric
effect and quantum phenomena in general,
while the analogy of spreading waves, also
within everyone's experience, is again
ready to suggest that the undulatory theory
is incompatible with the facts to be explained.

Exact mathematical analysis, as we have
seen, establishes the falsity of this last sug-
gestion. On the other hand, the analogy
of hydrodynamical sources and sinks, which
do not come within our everyday experience, suggests the possibility of explaining
the facts in terms of a pure undulatory
theory, and has nothing to fear from exact
mathematical analysis, from which indeed
it receives nothing but encouragement. As
the phenomena of everyday life do not
take place under conditions of thermodynamical equilibrium, they are not reversible processes in the thermodynamical
sense, whence it results that common
events cannot provide analogies for reversible processes; all the waves we ever meet
diverge and we have practically no direct
experience of convergent waves. Being
unable to find an everyday analogy to the
photoelectric effect in terms of waves, we
have perhaps jumped too hastily to the
conclusion that the effect cannot be explained in terms of waves, and have then
been tempted to rush to explanations in
terms of bullet-like atoms of radiation, for

which analogies abound. A great number
of such explanations have been suggested,
some of them of great ingenuity, but all,
so far as I know, meet their doom in the
established facts of interference and diffraction.

Again the photoelectric effect, like the
structure of the hydrogen atom, reveals the
inadequacy of the classical electrodynamics.
According to the model we have just con-
structed for the explanation of the photoelectric effect in terms of the classical laws,
the periods of motion in the absorbing and
emitting atoms are the same, and all space
is filled with radiation of precisely these
periods. We can describe the action of
the model by saying that one atom emits
radiation of the same periods as its own
motions, while the other atom absorbs this
radiation by resonance. But in Nature it is
only for radiations of very low frequency
that any period in the atom coincides with
that of the radiation which it emits or absorbs. We may think of the atom as acting
as a "source" for radiations of certain
definite frequencies, and as a "sink" for

radiations of other definite frequencies, but as a rule the atom does not beat time to any of these frequencies. Only in the extreme case of excessive slowness do the "source" and "sink" actions reduce to ordinary emission and absorption by resonance; in the more general case, I know of no phenomenon in the whole of physics which helps us in the least to comprehend the physical processes at work. With this fact before us, not much meditation is needed to convince us that we are still very far from understanding the working of the atom or the true meaning of atomicity and quanta.

Printed in the United States
By Bookmasters